Scrapie Disease in Sh